Leaping Learners
Education, LLC

For more information and resources visit us at:
www.leapinglearnersed.com

Published by Leaping Learners Education, LLC, Verona, NJ 07044 .

Every attempt has been made to credit each photo. Please contact us if there has been an error and we will resolve the issue.

Photo Credits
Fish diver graphics © Alexey Bannykh/stock.adobe.com

Cover, © BlueOrange Studio/shutterstock.com, ; Pg. 1-2 © Edgieus/shutterstock.com; Pg. 3-4 © max_776/stock.adobe.com; Pg. 5-6 © HelloRF Zcool/shutterstock.com; Pg. 7-8 © lophius/shutterstock.com; Pg. 9-10 © zaferkizilkaya/shutterstock.com; Pg. 11-12 © Damsea/shutterstock.com; Pg. 13-14 © Napa/shutterstock.com; Pg. 15-16 © H.Tanaka/shutterstock.com; Pg. 17 © Rich Carey/shutterstock.com; Pg. 18 © Rich Carey/shutterstock.com; Pg. 20 © Microgen/stock.adobe.com; Pg. 22 ©jannoon028/shutterstock.com; Pg. 23 © nikiteev_konstantin/shutterstock.com

All designs by Sean Bulger
All other pictures by Sean Bulger or royalty-free from Pixabay.com

ISBN
978-1-948569-24-8

Dear Parents and Guardians,

Thank you for purchasing a *Fay Learns About* series book! After teaching students from kindergarten to second grade for more than seven years, I became frustrated by the lack of engaging books my students could read independently. To help my students engage with nonfiction topics, my wife and I decided to write nonfiction books for children. We hope to inspire young children to learn about the natural world.

Here at Leaping Learners Education, LLC, we have three main goals:

1. Spark young readers' curiosity about the natural world
2. Develop critical independent reading skills at an early age
3. Develop reading comprehension skills before and after reading

We hope your child enjoys learning with Fay. If you or your children are interested in a topic we have not written about yet, send us an email with your topic, and maybe your book will be next!

Thank you,

Sean Bulger, Ed.M

www.leapinglearnersed.com

Reading Suggestions:

Before reading this book, encourage your children to do a "picture walk," where they skim through the book and look at the pictures to help them think about what they already know about the topic. Thinking about what they already know helps children get excited about learning more facts and begin reading with some confidence.

Preview any new vocabulary words with your child. Key vocabulary words are found on the last few pages of the book. Have your children use each new phrase in their own words to see if they understand the definition.

After previewing the book, encourage your child to read the book independently more than once. After they have read it, ask them specific questions related to the information in the book. Encourage them to go back and reread the relevant section in the book to retrieve the answer in case they forgot the facts.

Finally, see if your child can complete the reading comprehension exercises at the end of the book to strengthen their understanding of the topic!

This book is best for ages 6-8
but. . .
Please be mindful that reading levels are a guide and vary depending on a child's skills and needs.

Fay Learns About . . . Jellyfish

Written by Sean and Anicia Bulger

Table of Contents

Introduction.....1

Habitat.....3

Body.....5

Diet.....13

Predators.....17

Baby Jellyfish.....19

Danger.....21

Fun Facts.....23

Glossary.....24

Activities.....25

Hi! My name is Fay. I love to discover and learn new things. In this book, we will learn about jellyfish. Let's go!

Introduction

Have you ever seen a jellyfish? They are one of the oldest living animals on earth!

Habitat

Jellyfish live in all the world's oceans. They live in hot, cold, deep, and shallow parts of oceans.

EUROPE
ASIA
AFRICA
PACIFIC OCEAN
AUSTRALIA
ATLANTIC OCEAN

Body

Jellyfish do not have any bones or brains. Their bodies are soft and jelly-like.

Jellyfish do not have blood or a heart. Their bodies are mostly made of water.

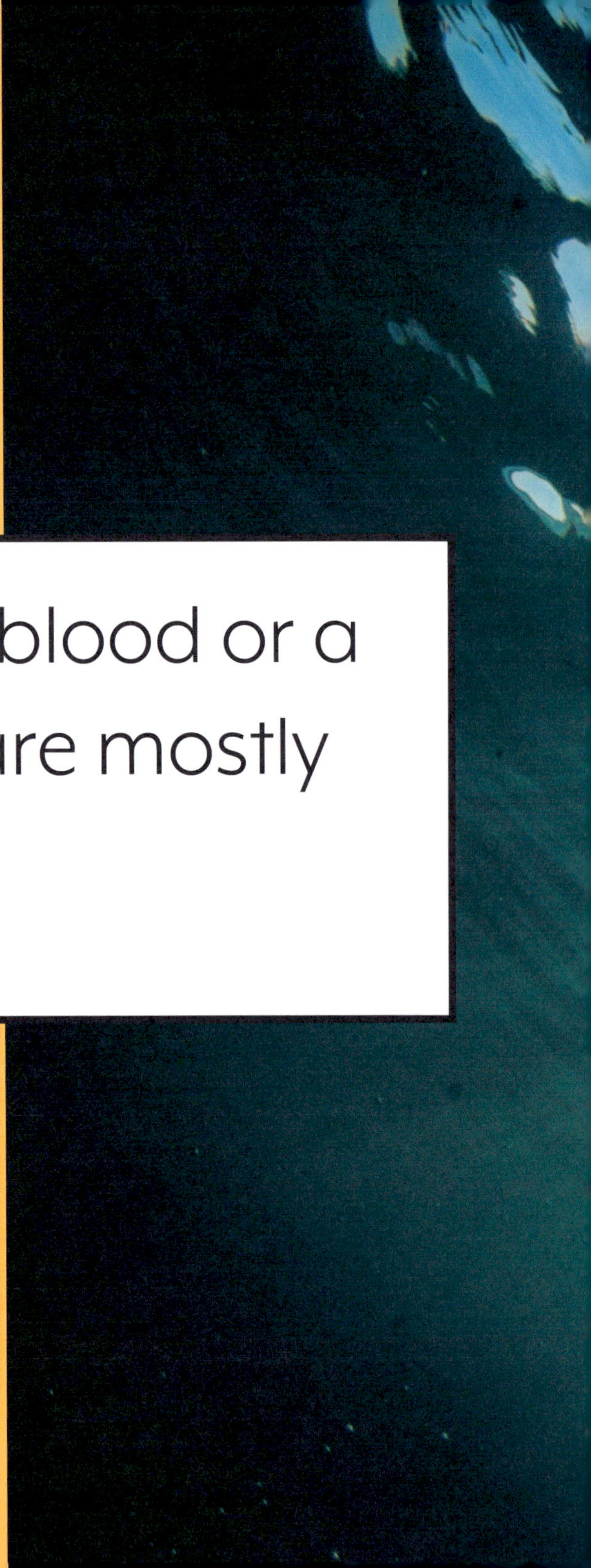

A jellyfish's body can come in many different colors. Some can even glow in the dark!

Jellyfish have an umbrella-shaped body with **tentacles** and a mouth in the middle of their body.

They move by squirting water out of their mouth to move forward.

Tricky word:
say...
um-brel-la
Mouth
Body
tentacles

Diet

Jellyfish eat small fish, shrimp, crabs, and tiny plants.

Jellyfish have long **stingers** on their tentacles to **paralyze** their prey.

After they paralyze their prey, jellyfish eat it with a mouth in the middle of their body.

PG 16

Predators

Sea turtles are jellyfish's main **predator**. They will quickly eat a floating jellyfish.

Some fish also eat jellyfish.

Baby Jellyfish

Most jellyfish lay eggs. Baby jellyfish called **polyps** are born from the eggs.

After they are born, they grow to become a jellyfish and live on their own.

Polyp

Danger!

While jellyfish do not attack people, they can sting people by accident.

Some jellyfish, such as the box jellyfish, can be very painful for people if they are stung.

WARNING
BEWARE OF JELLYFISH

The Lion's Mane jellyfish can be 120 feet long!

Jellyfish like the taste of peanut butter.

Some people eat jellyfish.

Glossary

A glossary tells the reader the meaning of important words.

Tentacles – A thin, flexible body part

Stingers – Toxic body part of an animal

Paralyze – To make somebody or something unable to move

Predator – An animal that eats other animals

Polyp – A jellyfish's second stage of life after hatching from an egg

Draw a picture of a jellyfish!

Choose 3 words from the glossary and write a sentence for each one.

1. ____

____

____

2. ____

____

____

3. ____

____

____

Quiz

1. True or false: Jellyfish have brains.

a. True

b. False

2. Which word best describes the body of a jellyfish?

a. Soft

b. Hard

c. Rough

3. What is the main idea of the section called “Danger”?

a. Why jellyfish are in danger

b. How jellyfish can be dangerous

c. What to do if you see a jellyfish

Common core standards:

RI. 2. 1 - Questions 1, 3

RI. 2. 4 - Question 2

4. What do jellyfish eat?

a. Sea Turtles

b. Sharks

c. Small fish

5. What does the word paralyze mean?

a. To make something unable to move

b. To sting something

c. To eat something

6. What do the labels on page 12 teach you?

a. Body parts of a jellyfish

b. How jellyfish swim

c. Where jellyfish live

Common core standards:
RI. 2. 1 - Question 1
RI. 2. 4 - Question 5
RI. 2. 2 - Question 4

Answer the following questions using information from the text:

1. Describe the body of a jellyfish.

2. How do jellyfish move in water?

3. What do jellyfish eat?

4. How are jellyfish born?

Want to learn about colors? Check out the "Clayton Teaches You About..." Series!

Great for early readers ages 4-6

Want to learn about Farm Animals? Check out the "Katie Teaches you About..." Series!

Great for early readers ages 4-6

Made in the USA
Las Vegas, NV
09 November 2026

57757233R00026